Patterns in the Sky

Lessons 1-2

Lesson 1
What Planets Orbit the Sun? 2

Lesson 2
What Patterns Do Earth and the Sun Follow? 12

Visit *The Learning Site!*
www.harcourtschool.com

Lesson 1

What Planets Orbit the Sun?

VOCABULARY
planet
orbit
telescope
moon
solar system

A **planet** is a large object in space that travels around a star. This is the planet Saturn.

To travel around an object is to **orbit**. This planet is orbiting the sun.

A **telescope** is a tool that scientists use to see objects that are far away. Objects seem closer and larger when you look at them through a telescope.

The **solar system** is the sun and everything that orbits it. The solar system includes nine planets and the planets' moons.

A **moon** is a large object that orbits a planet. The Earth has one moon that orbits it.

READING FOCUS SKILL
COMPARE AND CONTRAST

You **compare** things by looking for ways they are alike. You **contrast** things by looking for ways they are different.

Compare and **contrast** the planets. Find details that show how the planets are alike and different.

Planets and Moons

We live on Earth. Earth is a planet. A **planet** is a large body of rock or gas in space. Planets are moving. Earth is moving right now, but you cannot feel it. Planets **orbit**, or travel around, a star. Earth is one of nine planets that orbit the sun.

It takes Earth one year to orbit the sun one time.

The moon is smaller than Earth.

A **moon** is a large body that orbits a planet. Earth has one moon that orbits it. The moon takes about one month to orbit Earth once. You can see the moon in the sky at night.

The Solar System

The **solar system** is the sun and the things that orbit it. The sun is the center of the solar system. The word *solar* means "sun."

There are nine planets in the solar system. The moons that orbit the planets are part of the solar system, too.

Focus Skill **How do the planets move in the solar system?**

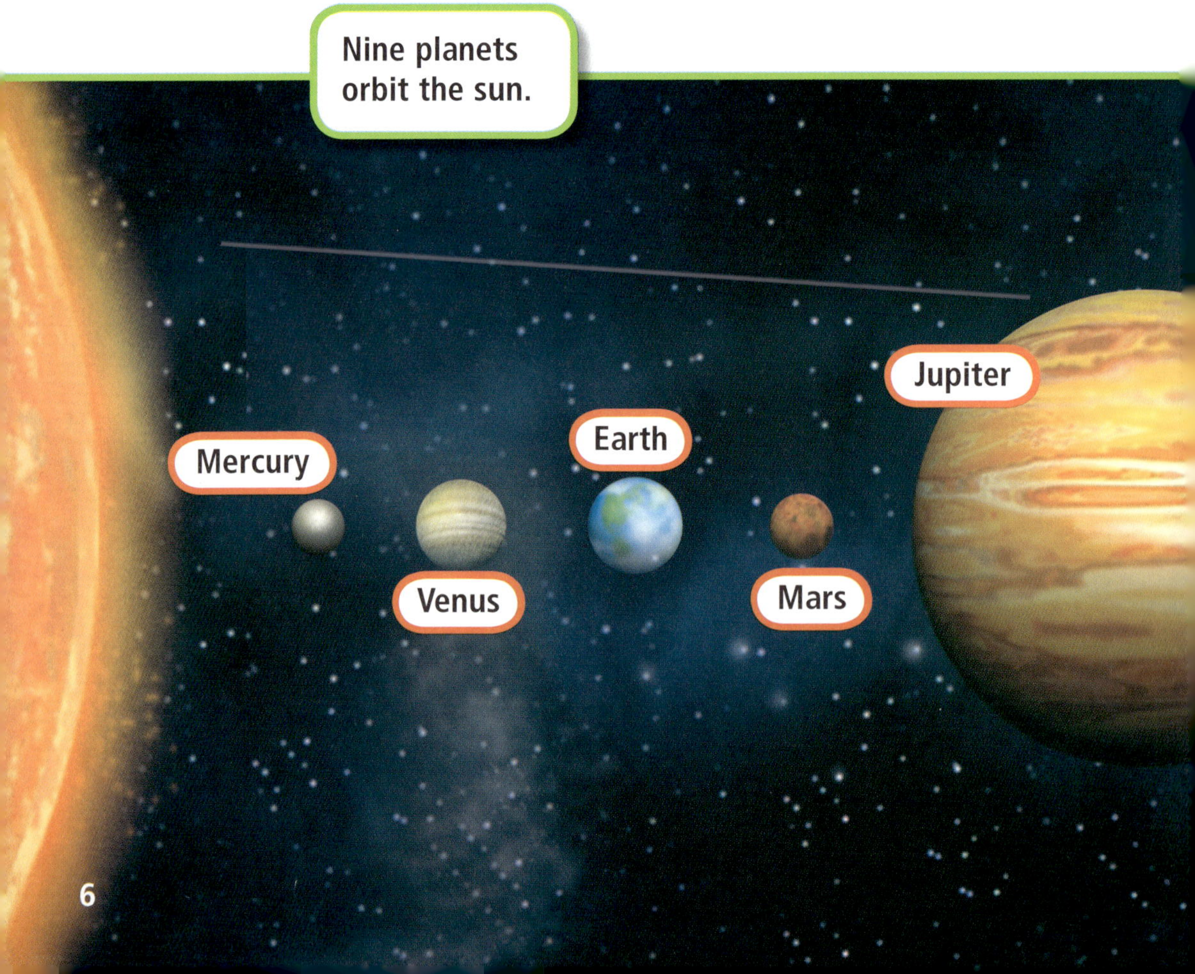
Nine planets
orbit the sun.
Mercury
Venus
Earth
Mars
Jupiter

Looking at the Sky

A **telescope** is a tool that helps you see things that are far away. Telescopes make things look closer and larger. A telescope helps you see details. You need a telescope to see the planets that are far away.

Focus Skill **How is looking through a telescope different from using your eyes alone?**

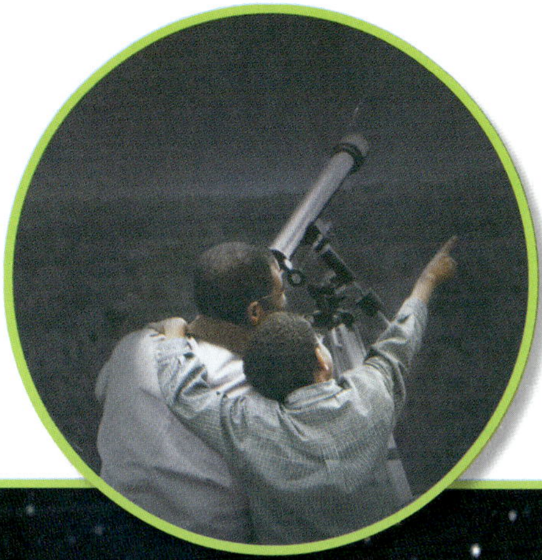

These two people are using a telescope. ▶

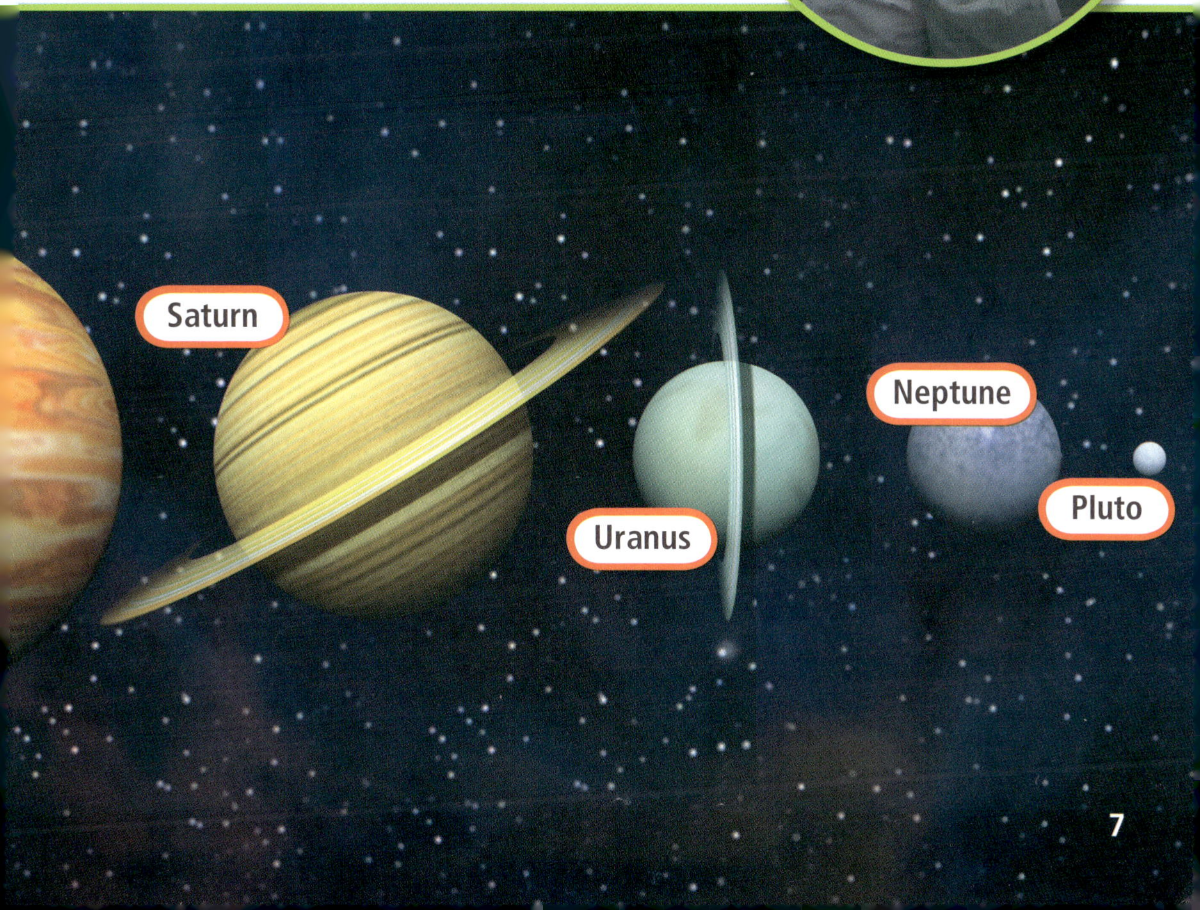
Saturn
Uranus
Neptune
Pluto

The Inner Planets

There are nine planets in the solar system. The four planets that are closest to the sun are called the inner planets. The inner planets are Mercury, Venus, Earth, and Mars.

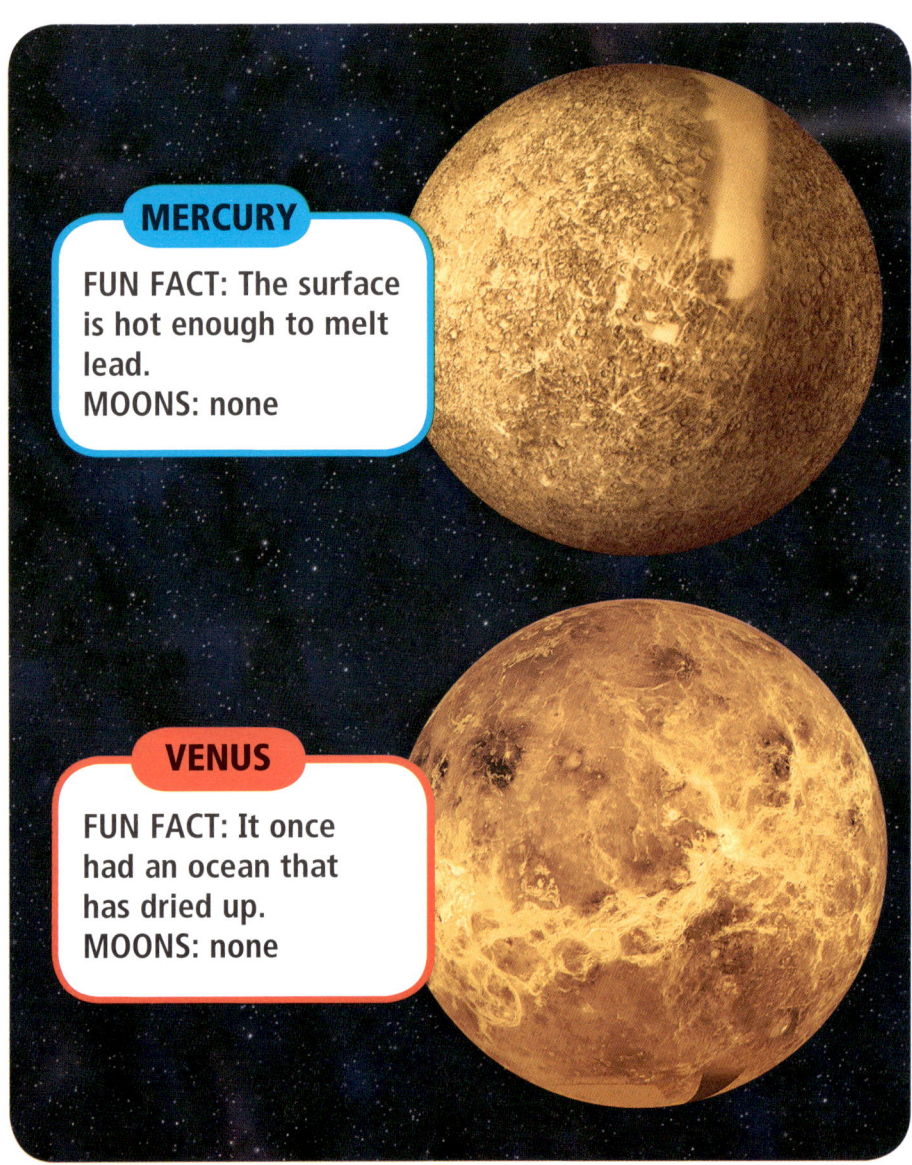

MERCURY
FUN FACT: The surface is hot enough to melt lead.
MOONS: none

VENUS
FUN FACT: It once had an ocean that has dried up.
MOONS: none

EARTH

FUN FACT: Earth is the only planet with water.
MOONS: 1

All the inner planets have rocky surfaces. They are also warmer than some of the other planets because they are closer to the sun.

 How is Earth different from other inner planets?

MARS

FUN FACT: Mars has the largest volcano in the solar system.
MOONS: 2

The Outer Planets

The five planets that are farthest from the sun are called the outer planets. They are Jupiter, Saturn, Uranus, Neptune, and Pluto. Most of them are larger than the inner planets. The outer planets are made of frozen gases.

 What is one way the outer planets are different from the inner planets?

JUPITER
FUN FACT: It has the Great Red Spot, which is a storm.
MOONS: more than 60

SATURN
FUN FACT: Saturn has many rings around it.
MOONS: more than 45

URANUS
FUN FACT: Uranus spins on its side.
MOONS: more than 25

NEPTUNE

FUN FACT: When Pluto crosses Neptune's orbit, it will be the planet farthest from the sun.
MOONS: more than 10

PLUTO

FUN FACT: Pluto is made of ice.
MOONS: 1

Review

Focus Skill

Complete these sentences to compare the planets.

1. Nine planets _____ the sun.

2. All _____ are large bodies of rock or gas that travel around the sun.

Complete these sentences that contrast the planets.

3. The _____ planets are closer to the sun than the _____ planets.

4. Some planets have one or more _____ orbiting them.

Lesson 2

VOCABULARY
rotation
axis
Northern Hemisphere

What Patterns Do Earth and the Sun Follow?

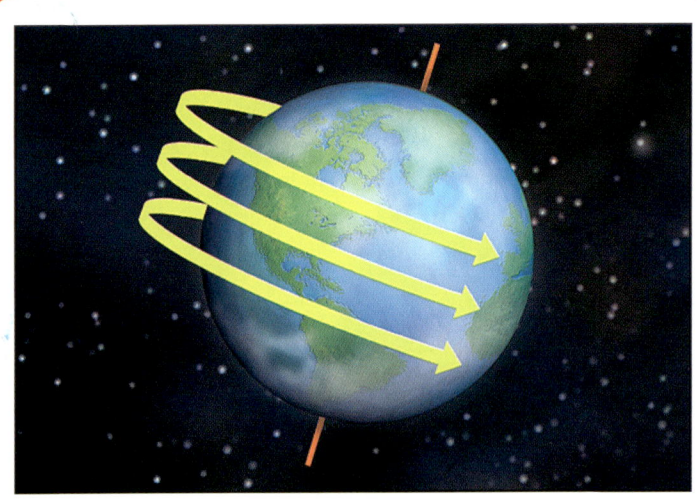

The Earth is always spinning. The spinning is called **rotation**. One rotation of Earth takes 24 hours.

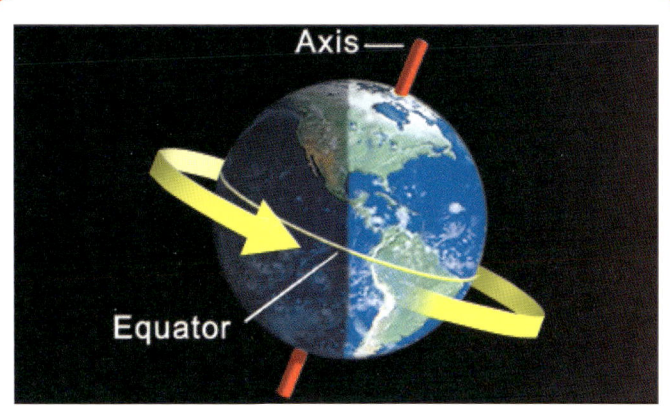

Earth has an imaginary line that runs through it. The line runs from the North Pole to the South Pole. We call this the Earth's **axis**. Earth spins around its axis.

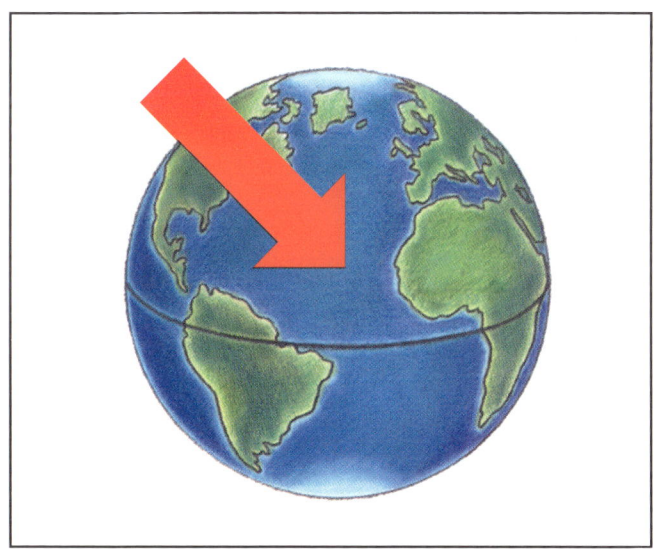

The **Northern Hemisphere** is the northern half of Earth.

> **READING FOCUS SKILL**
> **CAUSE AND EFFECT**
>
> A **cause** is something that makes another thing happen. An **effect** is the thing that happens.
>
> Look for the effects of the movements of Earth.

Day and Night

Earth moves in two ways. **Rotation** is the spinning of Earth. It takes Earth 24 hours, or one day, to make one rotation.

Earth also orbits the sun. It takes Earth 365 days to orbit the sun.

Earth is rotating, or spinning.

The sun is lighting only the side of Earth that is facing it. This is why we have day and night. The part of Earth that is lit by the sun is always changing because of Earth's rotation. When it is daytime for one half of Earth, it is night for the other half.

Focus Skill **What causes day and night?**

◀ The sun is shining on California.

At the same time, it is night on the other half of the Earth. ▶

Shadows Change

Every day, the sun comes up in the east. Later, the sun sets in the west. Notice how the sun's position changes in the pictures on the next page. Shadows change because the sun's position changes.

 What causes shadows to get longer and shorter?

The sun is rising. Think about how the sun will move across the sky.

The shadows show us how this part of Earth has rotated away from the sun. ▶

The Sun's Positions

Earth has an imaginary line that runs through it. The line is called its **axis**. Earth's axis is not straight up and down. It is tilted toward or away from the sun. The part of Earth that is tilted toward the sun has summer. The part that is tilted away from the sun has winter.

California is in the **Northern Hemisphere**, the half of Earth that is closer to the North Pole. In summer, the sun is higher in the sky, so the days are longer. In winter, the sun is lower in the sky and the days are shorter.

In summer, the sun is higher in the sky. ▼

▲ In winter, the sun is lower in the sky.

The Northern Hemisphere is tilted away from the sun. It has winter. ▶

◀ The Northern Hemisphere is tilted toward the sun. It has summer.

Review

Fill in these cause and effect statements.

1. The sun seems to move across the sky because of the _____ of Earth.

2. The tilt of Earth's _____ causes winter and summer.

Fill in the missing term.

3. California is in the _____ _____ . This is the northern half of Earth.

4. Earth _____ the sun. This takes 365 days.

GLOSSARY

axis (AK•sis) An imaginary line that runs through Earth from the North Pole to the South Pole

moon (MOON) A large body that orbits a planet

Northern Hemisphere (NAWR•thern HEM•ih•sfir) The northern half of Earth

orbit (AWR•bit) To travel around an object

planet (PLAN•it) A large body of rock or gas in space that orbits a star

rotation (roh•TAY•shuhn) The spinning of Earth on its axis

solar system (SOH•ler SIS•tuhm) The sun and objects that orbit it, including the planets and their moons

telescope (TEL•uh•skohp) A tool that makes faraway objects seem closer and larger